RAND NATIONAL DEFENSE RESEARCH INSTITUTE

Penaid Nonproliferation

Hindering the Spread of Countermeasures Against Ballistic Missile Defenses

Richard H. Speier, K. Scott McMahon, George Nacouzi

T0308348

Prepared for the Naval Postgraduate School, Project on Advanced Systems and Concepts for Combating WMD

This research was sponsored by the Defense Threat Reduction Agency and conducted within the International Security and Defense Policy Center of the RAND National Defense Research Institute, a federally funded research and development center sponsored by the Office of the Secretary of Defense, the Joint Staff, the Unified Combatant Commands, the Navy, the Marine Corps, the defense agencies, and the defense Intelligence Community under Contract W74V8H-06-C-0002.

Library of Congress Cataloging-in-Publication Data

ISBN: 978-0-8330-8149-0

The RAND Corporation is a nonprofit institution that helps improve policy and decisionmaking through research and analysis. RAND's publications do not necessarily reflect the opinions of its research clients and sponsors.

Support RAND—make a tax-deductible charitable contribution at www.rand.org/giving/contribute.html

RAND® is a registered trademark.

RAND OFFICES
SANTA MONICA, CA • WASHINGTON, DC
PITTSBURGH, PA • NEW ORLEANS, LA • JACKSON, MS • BOSTON, MA
DOHA, QA • CAMBRIDGE, UK • BRUSSELS, BE
www.rand.org

Preface

The proliferation of weapons of mass destruction (WMD) becomes a greater threat when accompanied by the proliferation of effective means of delivery. Proliferator nations are acquiring the means of delivery—most dramatically, ballistic missiles.

Defenses against ballistic missiles are accepted instruments for dealing with missile proliferation. Such defenses can help protect friends and allies from missile attack. The prospect of such protection can reduce the incentive for potential proliferators to acquire WMD and their ballistic missile delivery systems. Once proliferation has occurred, missile defenses can reduce the expected effects of proliferators' forces and thus help deter aggression.

These benefits will be lost, or at least reduced, if proliferators can acquire effective countermeasures against missile defenses. Such countermeasures, when incorporated in an attacker's missile, are known as penetration aids, referred to here as *penaids*. The subject of this documented briefing is an approach to hindering the proliferation of penaids.

This documented briefing was prepared in 2012–2013 under the Naval Postgraduate School research task "Penaid Nonproliferation: New Measures to Dissuade WMD Proliferation and Reinforce Deterrence." It should be of interest to individuals and organizations concerned with missile defense and with missile and WMD nonproliferation.

This research was sponsored by the Defense Threat Reduction Agency and conducted within the International Security and Defense Policy Center of the RAND National Defense Research Institute, a federally funded research and development center sponsored by the Office of the Secretary of Defense, the Joint Staff, the Unified Combatant Commands, the Navy, the Marine Corps, the defense agencies, and the defense Intelligence Community.

For more information on the International Security and Defense Policy Center, see http://www.rand.org/nsrd/ndri/centers/isdp.html or contact the director (contact information is provided on the web page).

Contents

Summary

This research describes an approach to hindering the spread of countermeasures against ballistic missile defenses. (Such countermeasures, when incorporated in an attacker's missile, are also called penetration aids, or *penaids*.) The approach involved compiling an unclassified list of penaid-relevant items that might be subject to internationally agreed-upon export controls.

The list is formatted to fit into the export-control structure of current international policy against the proliferation of missiles capable of delivering weapons of mass destruction. This policy, the Missile Technology Control Regime, creates two levels of control. One is a set of tight restrictions against a small number of items, such as complete missiles or their major subsystems. The other is a set of case-by-case export reviews for lower-level components and dual-use items.

This report recommends controls on 19 penaid-relevant items. More specifically, it recommends the tightest controls on three of those items: complete, integrated countermeasure subsystems; complete subsystems for missile defense test targets; and boost-glide vehicles. It offers as candidates for the tightest controls ten other items, such as re-entry vehicle replicas or decoys. But because these ten items are not complete subsystems, it identifies the possibility of treating them to a case-by-case review to improve the negotiability of the controls. Finally, the report identifies six classes of items, including test facilities and equipment, that could appropriately be subject to case-by-case review because of their utility for other applications, such as peaceful satellites.

Acknowledgments

Sponsors and Interviews

- Defense Threat Reduction Agency
- Naval Postgraduate School, Project on Advanced Systems and Concepts for Combating WMD
- Agencies and contractors interviewed

  - OSD
  - DoS
  - DoC
  - MDA
  - DTSA
  - MIT Lincoln Lab
  - Vision Centric Inc.
  - Gomez Research

  - deciBel Research
  - SMDC
  - MSIC
  - ODNI
  - CIA
  - NASIC
  - Sandia Lab
  - Bruce Haselman

RAND 1

The funds for this research were provided by the Defense Threat Reduction Agency (DTRA) and administered by the Naval Postgraduate School. Special recognition goes to David Hamon, formerly of DTRA, Anne Clunan of the Naval Postgraduate School's Project on Advanced Systems and Concepts for Combating WMD, and Meghan Rasmussen, also at the Naval Postgraduate School, for making this project a reality.

More than three dozen individuals from the organizations listed in slide 1 provided guidance for this research. The organizations were the Office of the Secretary of Defense, the U.S. Department of State, the U.S. Department of Commerce, the Missile Defense Agency, the Defense Technology Security Administration, the Massachusetts Institute of Technology's (MIT's) Lincoln Laboratory, Vision Centric Inc., Gomez Research, deciBel Research, the U.S. Army's Space and Missile Defense Command, the Missile and Space Intelligence Center, the

Office of the Director of National Intelligence, the Central Intelligence Agency, the U.S. Air Force's National Air and Space Intelligence Center, the U.S. Department of Energy's Sandia Laboratory, and the individual Bruce Haselman.

Individuals from more than half of the participating organizations attended a July 25, 2012, workshop at RAND to review and comment on preliminary findings. With the workshop comments in hand, the authors made extensive revisions and expansions to a draft of this report. Particular credit is due to Allen Dors at MIT's Lincoln Lab and Brian Chow at RAND for their insightful reviews of multiple drafts. However, this report reflects the views of its RAND authors and not necessarily those of any other individuals or any organization.

CHAPTER ONE

Penaid Nonproliferation

Research Objective

Assist U.S. agencies in hindering the spread of countermeasures against ballistic missile defenses by developing an unclassified list of penaid-relevant items that might be subject to internationally agreed-upon export controls

RAND   2

This research was designed to assist U.S. agencies charged with generating policies to discourage the proliferation of weapons of mass destruction (WMD) and ballistic missile delivery systems, thereby strengthening deterrence. The objective was to develop new measures to restrict the proliferation of countermeasures (also known, when incorporated in an attacker's missile, as penetration aids, or *penaids*) against ballistic missile defenses. It is necessary to identify the science and technology underpinning the development of penaids before policies can be designed to control the threat. Therefore, the study team focused on answering the following overarching research question: What technologies and equipment, if proliferated, would constitute an emerging penaid threat to the United States?

The scope was limited to countermeasures against ballistic missile defenses. Countermeasures against cruise missile defenses involve substantially different technologies and will be investigated in a future study.

Research Method

Conduct a literature review, interviews, a technical assessment, and a subject-matter expert workshop to generate a list of penaid-relevant items that might be subject to internationally agreed-upon export controls

RAND   **3**

The RAND National Defense Research Institute drew on its expertise in the several subjects relevant to the project: U.S. ballistic missile defense systems; domestic and foreign development of penetration aids and related technology and equipment; relevant U.S. aerospace systems, technologies, and industry; and related proliferation/nonproliferation matters. RAND analysts conducted a literature review and interviews to identify data sources and leading government and nongovernment experts on subjects relevant to the project. The study team conducted structured interviews and an independent technical assessment to develop a preliminary characterization of the technologies and equipment most critical to the emerging penaid threat. Thereafter, the team invited a selected group of experts to participate in a one-day workshop to review the initial characterization of penaid technologies and equipment.

Findings by a Congressional Commission

"The United States should . . . develop effective capabilities to defend against increasingly complex missile threats . . . [which] include technologies intended to help in-coming missiles penetrate the defense (so-called penetration aids). . . . The United States should also work with Russia and China to control advanced missile technology transfer."

Source: Congressional Commission on the Strategic Posture of the United States (known as the Perry/Schlesinger Commission), *America's Strategic Posture: The Final Report of the Congressional Commission on the Strategic Posture of the United States,* Washington, D.C.: United States Institute of Peace Press, May 2009, http://media.usip.org/reports/strat_posture_report.pdf.

RAND   4

The project was suggested in 2009 by a commission chartered by Congress and headed by two former Secretaries of Defense. The commission recommended a focus on technology transfers from Russia and China—an approach referenced in slide 4 and consistent with the broader international scope developed here.

In fact, the commission's report represented a turning point in interest in penaid nonproliferation. According to the recollection of a RAND study team member, when international missile nonproliferation controls were being designed in 1982–1983, the U.S. participants considered including controls on penetration aids. However, doing so would have required bringing in additional expertise, complicating the international discussions, and it would have addressed a problem that was likely to be decades away. Thus, the matter was deferred.

The issue arose again almost two decades later—in a 2000 contractor study for the Office of the Secretary of Defense (see McMahon, Orman, and Speier, 2000). However, Missile Defense Agency concerns deferred action on the matter again.

In 2007, one of the authors of this RAND report published an article on penaid nonproliferation (see Speier, 2007). The article recommended export-control modifications similar to those discussed here. However, the article was based on a more limited field of interviews and lacked the benefit of the 2012 RAND workshop, which brought together experts to comment on each others' positions and, consequently, to modify them. The 2007 report did not include some of the most important items recommended here for the tightest export controls, it formulated items differently from the more operationally appropriate descriptions here, and it did not fit into the structure of export controls as closely as the suggestions in this report.

Soon after the 2009 report of the congressional commission, the Defense Threat Reduction Agency (DTRA) began discussions with RAND that led to the research described here.

Policy Refinements to the Objective

- **Fullest possible list**

- **MTCR format**

- **Technology, not policy**

- **Be specific or be general**

- **Avoid "specially designed"**

RAND   5

Interviewees for this research suggested several refinements to the control list. One was to develop a broad list of penaid-related items from which the most important might later be selected.

Another was to put these items in the format of the Missile Technology Control Regime (MTCR), the international instrument for hindering the spread of WMD-capable missiles. The MTCR Annex is a control list of items presented in a format usable by government export-control officials (see MTCR, 2012).

A third suggestion was to focus exclusively on penaid-relevant technology, not on policy questions. This was easier said than done, however, because—as discussed later—the placement of items on the MTCR Annex determines the degree to which the export of such items will be restricted. Some interviewees advised the RAND researchers that it would help export control and customs officials if they were as specific as possible in identifying candidate items for controls. Others suggested a more general approach to avoid unintended information transfer to proliferators. As a result, the study team had to choose an approach that would take into account both requests.

Finally, the international export control community is attempting to minimize the use of the term *specially designed* in identifying items to be controlled. The MTCR defines the term as follows:

> "Specially Designed" describes equipment, parts, components or software which, as a result
> of "development," have unique properties that distinguish them for certain predetermined

purposes. For example, a piece of equipment that is "specially designed" for use in a missile will only be considered so if it has no other function or use. Similarly, a piece of manufacturing equipment that is "specially designed" to produce a certain type of component will only be considered as such if it is not capable of producing other types of components. (MTCR, 2012, p. 15)

The international export control community's concern is that the term *specially designed* describes intent rather than the physical features of a controlled item. However, in some cases, RAND researchers found no alternative for the phrase.

Penaids: There's a Lot Out There

Google   "penetration aids"   🔍

Web   Images   Maps   Shopping   More ▾   Search tools

About 28,000 results (0.31 seconds)

Penetration aid - Wikipedia, the free encyclopedia
en.wikipedia.org/wiki/Penetration_aid
A penetration aid is a device or tactic used to increase an intercontinental ballistic
missile (ICBM) warhead's chances of penetrating a target's defenses.

penetration aids - The Free Dictionary
www.thefreedictionary.com/penetration+aids
Techniques and/or devices employed by offensive aerospace weapon systems to
increase the probability of penetration of enemy defenses. Want to thank TFD ...

Missile penetration aids
fractalsoftworks.com › ... › Starsector › Suggestions
10 posts - 7 authors - Mar 10, 2012
Seems I'm not alone in my dissatisfaction with missiles. Certainly, one CAN use them at
the right time, but "the right time" is a pretty narrow case ...

Source: Google.

RAND   6

There is a great deal of openly available information on penaids. Entering "penetration aids" (in quotes) in the Google search engine yields thousands of hits with hundreds of nonrepeating hits. "Penaids," a more colloquial term, is also well represented online.

Penaids: There's a Lot Out There (Cont'd)

Penetration aid

From Wikipedia, the free encyclopedia

A **penetration aid** is a device or tactic used to increase an intercontinental ballistic missile (ICBM) warhead's chances of penetrating a target's defenses. These can consist of both physical devices carried within the ICBM, as well as tactics that accompany its launch, and may include one or more of the following:

- the MIRV bus carrying the nuclear warheads can have some form of stealth technology, thereby hindering detection before the warhead reentry vehicles are released.
- chaff: Chaff wires may be deployed over a large area of space, creating a large, radar-reflecting object that will obscure incoming warheads from defensive radar.
- decoys: Decoys consist of mylar balloons that can be inflated in space and are designed to have the same radar characteristics as the warhead. Because the warhead and the decoy balloons may be at different temperatures, the warhead and the balloons may both be surrounded by heated shrouds that put them all at the same temperature. This defeats attempts to discriminate between decoys and warheads on the basis of temperature, which can confuse an enemy's missile defense systems.
- Incidental or deliberate fragmentation of the final-stage rocket booster can cloud the enemy's radar by projecting a radar cross-section much larger than the actual missile.[1]
- radar jammers: These are transmitters that can be deployed on the decoys and the warhead to jam the frequencies used by defensive radars.
- nuclear radar blackout: A nuclear device may be deliberately exploded in space by the

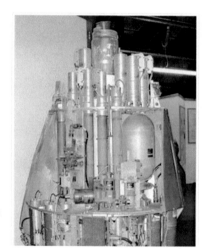

Source: "Penetration Aid," *Wikipedia*, last updated March 17, 2013, http://en.wikipedia.org/wiki/Penetration_aid (photo by Chris Gibson).

RAND   7

The first Google hit is a *Wikipedia* reference that displays an entire countermeasure sub-system, the British Chevaline Penetration Aid Carrier, as depicted in slide 7.

Penaids: Controls Will Help

- **Penaids ought to be**
  - **Matched to the defensive system**
  - **Integrated with the offensive system**
  - **Able to function in space and/or re-entry**
  - **Flight-tested with appropriate data collection**

- **Limited suppliers able to meet these criteria**

- **Without external assistance, development entails significant additional cost and time**

- **Missile performance may suffer**

RAND

8

In spite of the wide range of publicly available information on penaids, there is hope for the effectiveness of export controls. This is because, to be reliable, penaids must meet several difficult criteria:

- They must be matched to the defensive system they are intended to penetrate—meaning that details of the defensive system must be known and understood.
- They take volume, weight, and power away from the payload of the offensive system and must not interfere with the missile's functions or reliability—meaning that sophisticated systems engineering must integrate penaids with the rest of the payload.
- They must be able to survive the launch environment and function in their intended operational environment, i.e., space and in some cases re-entry—another engineering challenge.
- To be reliable, penaids must be tested in space and, in some cases, re-entry with appropriate instrumentation—a more difficult challenge than merely observing the arrival of a re-entry vehicle in a target area.

There are few suppliers who can help proliferators meet all these criteria. As noted earlier, a congressional commission identified Russia and China as key potential suppliers. Russia is a full participant in the MTCR. China is not, but it maintains that it adheres to an early version of the MTCR.

Without external assistance, penaid development can be costly and time-consuming. In the 1960s, the United States reportedly spent $300–400 million per year on penaid research and development (Sessler et al., 2000). The British Chevaline program, conducted largely during the 1970s with limited U.S. assistance, cost more than £1 billion and took more than a decade for research, development, and entry into service (Dommett, 2008; Jones, 2005; Panton, 2004a, 2004b, 2006).

The weight and other physical demands of penaids can reduce missile performance—more so if external assistance is limited. For example, the British Polaris missiles reportedly suffered range and payload penalties as a result of being equipped with the Chevaline subsystem (see "Chevaline," 2013; Simpson, 2004).

Recent reports by the U.S. Government Accountability Office describe difficulties with U.S. missile defense test targets, which contain penaid-like technology (GAO, 2008, 2011).[1] Problems with target performance have added at least $1 billion to the missile defense budget, and target failures and anomalies have "negatively affected many of the missile defense elements" (GAO, 2008, p. 4). A new target system, including launch vehicles, has cost more than $600 million for the first six targets. These costs and difficulties can be minimized with external assistance that provides proven penaid hardware or technology. But if such assistance is restricted by export controls, challenges and expenses can be expected to afflict the development of reliable penaids.

Given the difficulty of developing reliable penaids, one might ask whether a proliferator would settle for the development of less reliable penaids—perhaps by not testing them in space or by designing them in the absence of some critical information about the defensive system. As noted earlier, any penaid will take volume, weight, and power away from the lethal payload. Consequently, there is a performance trade-off to consider, possibly in terms of a range penalty or a reduction of payload lethality. This penalty must be weighed against a penaid subsystem of uncertain value. Moreover, a penaid—if not adequately tested—may interfere with the functioning of the missile system itself. These are significant considerations for a proliferator looking to develop penaids at low cost but high risk.

---

[1] See also slide 21 and surrounding text.

General Problems with a Penaid Control List

- **Potential tripling of MTCR Category I list**

- **Utility of some items for satellites, satellite defense, and missile defense itself**

- **Tension between being general or being specific**

- **Negotiability**

RAND

9

There are several broad problems with developing a list of items to be controlled.

The first concerns the MTCR's tight focus on a few items to be subject to the strictest export restrictions—the Category I list, to be discussed later. Only eight classes of such items are listed under Category I in the MTCR Annex. This study explored 19 additional items, potentially diffusing the MTCR's tight focus if all the items were added to Category I.

The second problem concerns the dual uses of many penaid-relevant items. Because they must function in space, some of the items are usable for satellites. And, as explained later, some are usable for missile defenses themselves.

We have already touched on the difficulty of defining an item with enough specificity that export-control officials can recognize it while retaining enough generality that the control list does not become a "to-do" list facilitating unintended information transfer to a proliferator.

With any international control list, international negotiability is a concern. By and large, the study team has deferred to the expertise of export-control officials to identify the best ways to address such concerns. This report does, however, point out alternative sites on the MTCR Annex for penaid-relevant items. The choice of a site may affect negotiability.

Candidates for a Control List

- **Complete penaids**
- **Test targets**
- **Boost-glide vehicles**
- **Canisters/dispensers**
- **Post-boost subsystems**
- **Replicas/decoys**
- **Electronic countermeasures**
- **Chaff/obscurants**
- **RV/decoy signature control**
- **Plume signature control**
- **Wake modification**
- **Maneuvering subsystems**
- **Submunitions**
- **Multiple-object deployment**
- **Inflation/assembly items**
- **Hardening**
- **Attack warning sensors**
- **Fly-along sensors**
- **Test facilities/equipment**

RAND   10

The list in slide 10 serves as a "table of contents" for this report: 19 penaid-relevant items that might be subject to international export controls. Before discussing each of these items individually, we point out the key features of the MTCR that affect the treatment of these 19 items.

Topics Not Addressed in This Report

- **MTCR history**

- **Countermeasures against missile defense infrastructure and communications**

- **Penaid suppliers**

- **Cruise missile penaids**

RAND   **11**

To keep this report focused, many related topics are not covered here.

Beyond the elements of the MTCR discussed in this report, there is a background of more than 25 years of successes and shortcomings. Some of this background may be found in the writings of Speier (1991, 1995, 2000, 2007) and in the transcript of an interview with Principal Deputy Assistant Secretary of State for International Security and Nonproliferation Vann Van Diepen (Davenport, Horner, and Kimball, 2012).

This report does not examine all of the tactics that might be used to defeat missile defenses. It focuses on ballistic missile countermeasures that might be subject to international controls, i.e., countermeasures carried by missiles and potentially controlled by the MTCR.

Except in a few illustrations, this report does not discuss firms or governments that might supply penaids. As noted previously, this report does not discuss penaids for cruise missiles. Those topics are the subjects of follow-on research.

CHAPTER TWO

The Missile Technology Control Regime

How the MTCR Works

- **Designed to prevent the proliferation of**
  - **Rockets or UAVs capable of delivering a 500 kg payload to a range of at least 300 km**

    or
  - **Any rocket or UAV intended to deliver WMD**
- **Two categories**
  - **Category I – items subject to a strong presumption of export denial**
  - **Category II – items subject to a case-by-case review and no-undercut rule**
- **Enforced by international cooperation and/or U.S. sanctions**

RAND   **12**

The MTCR seeks to hinder the spread of rockets and unmanned aerial vehicles (UAVs)—regardless of purpose (e.g., space launch, surveillance)—beyond a specified range-payload capability, or regardless of range-payload capability if intended to deliver WMD.

The MTCR's Category I list consists of eight classes of items subject to the tightest export restrictions. The MTCR guidelines state that such exports, if they occur at all, must be "rare" and subject to strong provisions with respect to supplier responsibility.

The MTCR's Category II list consists of items that can be used to make Category I items, as well as other missile items related to potential WMD delivery. However, Category II items are generally dual-use, applicable to purposes other than those related to Category I items. So, Category II exports are subject to greater flexibility but nevertheless require case-by-case export reviews and international procedures to avoid undercutting Category II denials by MTCR partners.

The MTCR has well-developed procedures for sharing export decision information among its members. The United States has legislation providing for sanctions against domestic and foreign entities that contribute to missile proliferation (see Speier, Chow, and Starr, 2001). In addition, there are United Nations Security Council sanctions, particularly against Iran and North Korea, that proscribe transfers of items in the MTCR Annex.

Key MTCR Items Considered in This Study

- **Category I:**
  - **Item 1 – delivery systems (500 kg/300 km)**
    - 1.A.1. **rockets**
    - 1.A.2. **unmanned aerial vehicles**
  - **Item 2 – complete subsystems for Item 1:**
    - 2.A.1.a. **individual stages**
    - 2.A.1.b. **RVs and their equipment**
    - 2.A.1.c. **rocket motors and engines**
    - 2.A.1.d. **guidance sets**
    - 2.A.1.e. **thrust vector controls**
    - 2.A.1.f. **safing, arming, fuzing, and firing mechanisms**

RAND   **13**

As noted in slide 13, the MTCR applies its restrictive Category I controls to two classes of complete delivery systems: rockets and UAVs above a specified capability threshold.

The MTCR also applies Category I controls to six classes of complete subsystems (Item 2.A set in slide 13) that can readily contribute to Category I delivery vehicles. Of these, some penaid-relevant items could be listed in an expansion of Item 2.A.1.b, which controls re-entry vehicles (RVs) and their equipment.

Key MTCR Items Considered in This Study (Cont'd)

- **Category II:**
  - **Item 4 – propellants and chemicals**
  - **Items 5, 7, and 8 – reserved for future use**
  - **Item 6 – structural materials**
  - **Item 10 – flight control**
  - **Item 11 – avionics**
  - **Item 15 – test facilities and equipment**
  - **Item 16 – modeling-simulation and design integration**
  - **Item 17 – stealth**
  - **Item 18 – nuclear effects protection**

RAND   **14**

The MTCR's less-restrictive Category II list consists of 15 classes of items and three open items ("reserved for future use"), which included hardware and technology that were later recategorized.

The Category II classes of items listed in slide 14, including the open items, are possible sites for listing some penaid items and their technology.

Items Also Covered by the MTCR

- **Test and production equipment**

- **Materials**

- **Software**

- **Technology**

RAND   **15**

Each hardware item listed in the MTCR Annex is generally accompanied by a list of related items shown in slide 15. In particular, design and production technology is treated at least as restrictively as the hardware item itself.

Penaid Subsystems Potentially Covered by the Current MTCR

• **Re-entry vehicles**

• **Stealth**

• **Nuclear effects protection**

RAND   **16**

It can be argued that the current (October 23, 2012) version of the MTCR Annex can be interpreted to cover some penaid technologies discussed in this report.

For example, in Category I, re-entry vehicles and their equipment are described as follows:

[Item 2.A.1.b] "Re-entry vehicles, and equipment designed or modified therefor . . [including] 1. Heat shields, and components therefor. . . 2. Heat sinks and components therefor . . . 3. Electronic equipment specially designed for re-entry vehicles." [The text represented by ellipses specifies usability for rockets with a 500 kg/300 km capability and exceptions for non-weapon payloads.]

On the one hand, it could be argued that any penaid subsystem carried by an RV is automatically covered by the current MTCR. But this is not obvious to some officials who implement the MTCR. And such phrases as "designed or modified therefor" leave open the question of whether "therefor" applies to equipment carried by the RV but not essential to its operation. In this report, we specify penaid items that need to be covered. If they are not clearly covered by the current MTCR, then it is worth considering whether to make their coverage explicit.

In Category II, some penaid technologies are arguably covered as stealth or nuclear effects protection. The relevant language is as follows:

[Item 17.A.1] Devices for reduced observables such as radar reflectivity, ultraviolet/infrared signatures and acoustic signatures (i.e. stealth technology). . . .

[Item 18] Nuclear effects protection . . . [specifically] 18.A.1 "Radiation Hardened" "micro-circuits" usable in protecting rocket systems . . . against nuclear effects (e.g. Electromagnetic Pulse [EMP], X-rays, combined blast and thermal effects) . . . 1 8.A.2. "Detectors" specially designed or modified to protect rocket systems . . . against nuclear effects . . . 18.A.3. Radomes designed to withstand a combined thermal shock greater than . . . accompanied by a peak over pressure of greater than . . . usable in protecting rocket systems . . . against nuclear effects. . . .

Once again, there is the question of making explicit those additional stealth items that need to be covered. And, with Item 18, there is currently no mention of penaid hardening technologies to protect against non-nuclear effects.

Early in the MTCR Annex, there is a series of definitions. In the annex, the definition of *payload* includes countermeasures equipment and post-boost vehicles. However, this definition does not constitute a control list. The purpose of the definition is to standardize the calculation of the mass delivered by a missile.

CHAPTER THREE

Items Proposed for Category I

Definitions

- **Penaid – In this report, countermeasures carried on an attacker's ballistic missile to defeat missile defenses**

- **Space-qualified – Capable of surviving the launch, space, and/or re-entry environments**

- **Electromagnetic spectrum – The full radiofrequency, infrared, optical, and ultraviolet spectra**

RAND   **17**

Slide 17 defines terms used frequently in this report. When these terms are used, they are meant with the definitions shown.

Penaids: Possible Category I Items

Missile-borne countermeasure subsystems, and penetration aids designed to saturate, confuse, evade, or suppress missile defenses and designed or modified for rocket systems capable of delivering at least a 500 kg payload to a range of at least 300 km, including . . .

RAND   18

The text in slide 18 applies to all candidate Category I items discussed in this report.

Complete Penaids

• **Complete, integrated, missile-borne countermeasure subsystems**

• **Issue**
  – **MTCR Item 2 (complete subsystems)**

RAND   **19**

The most obvious Category I candidate is a complete penaid subsystem. The obvious place for it on the MTCR Annex is Item 2, complete subsystems.

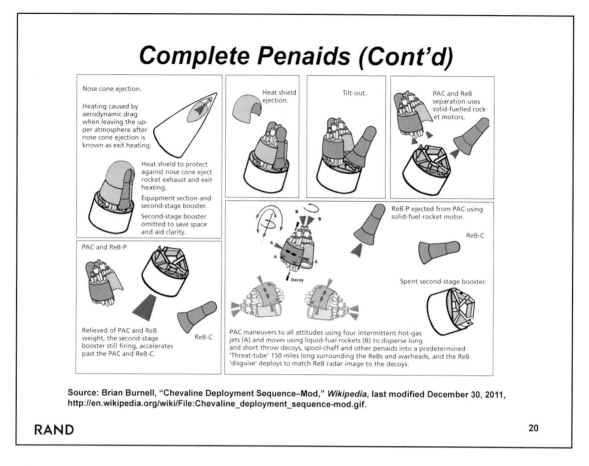

Complete Penaids (Cont'd)

Nose cone ejection.

Heating caused by aerodynamic drag when leaving the upper atmosphere after nose cone ejection is known as exit heating.

Heat shield to protect against nose cone eject rocket exhaust and exit heating.

Equipment section and second-stage booster.

Second-stage booster omitted to save space and aid clarity.

Heat shield ejection.

Tilt-out.

PAC and ReB separation uses solid-fuelled rocket motors.

PAC and ReB-P

ReB-C

Relieved of PAC and ReB weight, the second-stage booster still firing, accelerates past the PAC and ReB-C.

ReB-P ejected from PAC using solid-fuel rocket motor.

ReB-C

Spent second-stage booster.

PAC maneuvers to all attitudes using four intermittent hot-gas jets (A) and moves using liquid-fuel rockets (B) to disperse long and short-throw decoys, spool-chaff and other penaids into a predetermined 'Threat-tube' 150 miles long surrounding the ReBs and warheads, and the ReB 'disguise' deploys to match ReB radar image to the decoys.

Decoy

Source: Brian Burnell, "Chevaline Deployment Sequence–Mod," *Wikipedia*, last modified December 30, 2011, http://en.wikipedia.org/wiki/File:Chevaline_deployment_sequence-mod.gif.

RAND   20

This report includes illustrations—prepared by RAND or available online—of some of the items suggested for MTCR controls. For example, as depicted in slide 20, the British Chevaline countermeasure subsystem contains many components that must be operated in a strictly choreographed manner.

Test Targets

- **Complete subsystems for missile defense test targets designed to appear like RVs or countermeasures to missile defense sensors**

- **<u>Issues</u>**
  - **Affects missile defense cooperation. May require testing in supplier nation or supplier retaining jurisdiction or control of test targets until launch.**
  - **MTCR Item 2 (complete subsystems)**

RAND   **21**

Missile defense test targets simulate offensive missiles, RVs, or penaids. They are used in exercises of missile defense sensor and interceptor systems. Such targets create a number of proliferation problems. Their technologies may be indistinguishable from—or, at least, interchangeable with—those of penaids. At a minimum, their development and testing offer a perfect cover for the development and testing of penaids themselves. Consequently, they would appropriately be treated in the MTCR identically to complete penaids by being included in Item 2 of the MTCR Annex.

However, there is much legitimate international cooperation in missile defense. And missile defense capabilities do need to be tested against realistic targets. How can international cooperation occur if participants do not share test targets and their technology?

There are at least three possibilities for resolving this dilemma, all of which have precedents in nonproliferation practice. One is for nations receiving missile defenses to develop their own test targets. This has the disadvantage of permitting the development of penaid technology. A second possibility would involve allowing the recipient nation to conduct testing on the supplier's territory to avoid having the supplier export the test targets. This would have the additional advantage of not requiring the recipient of missile defense equipment to develop its own targets and infrastructure. A third possibility would have the supplier of test targets provide them under conditions permitted by the MTCR, with the supplier retaining jurisdiction or control of the items until they are launched. This alternative would prevent the test target or its technology from passing to the tester of the missile defenses while still allowing realistic testing on the recipient's territory.

Test Targets (Cont'd)
U.S. Test Targets

"L'Garde's first project was the Inflatable Exoatmospheric Object (IEO) in 1971. The most heavily instrumented inflatable target L'Garde has designed, tested, and flown was the Firefly. The firefly was instrumented to detect and change its state of motion through an on-board microprocessor and a L'Garde-developed lightweight coning-control system. Other targets and countermeasures objects include the (a) multi-balloon canister, (b) sounding rocket measurements program (SRMP) objects, (c) light replica decoy (LREP), (d) dual-decoy technology (DDT), and the (e) thrusted replica decoy (TREP)."

Source: L'Garde, Inc., "Missile Defense Targets and Countermeasures," web page, undated(b), http://www.lgarde.com/programs/missile-defense-targets-and-countermeasures. Used with permission.

RAND   22

Slide 22 provides another example of penaid-related information that is available online. L'Garde, Inc., was founded in 1971 in Orange County, California, to support ballistic missile defense through the development and manufacture of inflatable targets and decoy systems. It has exported some of these items.

Test Targets (Cont'd)
U.S. Test Targets

The Calibrated Orbiting Objects Project (COOP)

"The COOP was an MDA (Missile Defense Agency) supported project through AFRL [Air Force Research Laboratory]. Its objective was to develop calibrated orbiting objects for use in research into technologies necessary to build relatively inexpensive microsatellites for test and calibration of the U.S. ballistic missile defense system (BMDS) sensors. The first COOP was designed to simulate reentry vehicle motion and dynamics using the spacecraft attitude control system."

Source: L'Garde, Inc., "The Calibrated Orbiting Objects Project (COOP)," web page, undated(a), http://www.lgarde.com/programs/missile-defense-targets-and-countermeasures/coop. Used with permission.

RAND   23

As indicated in slide 23, the Calibrated Orbiting Objects Project developed test targets for the Missile Defense Agency. Test targets can be sophisticated, reproducing the behavior of RVs or decoys.

Boost-Glide Vehicles

- **Ballistically boosted glide vehicles designed to use aerodynamic forces to control their trajectory and their specially designed subsystems (including their thermal protection subsystems); to be treated as RVs and their subsystems**

- **Issues**
    - **Specially designed?**
    - **Do not include as Item 2.A.1.a (individual rocket stages) because not powered?**
    - **Include as note to Item 2.A.1.b (re-entry vehicles) "specially designed" – necessary in view of Item 2 note allowing Category II treatment for peaceful spacecraft?**

RAND   **24**

Unpowered boost-glide vehicles can function as RVs that do not follow a ballistic trajectory. For this reason, they are more difficult to track and intercept, and they can maneuver to defeat missile defenses. Although they may or may not leave the atmosphere—and thus may not strictly "re-enter"—they can appropriately be subject to the same treatment as RVs in the MTCR Annex. However, the re-entry of peaceful spacecraft is controlled by the same technologies, again raising the issue of "specially designed" or an exemption for peaceful spacecraft that already applies to RVs in the MTCR Annex.

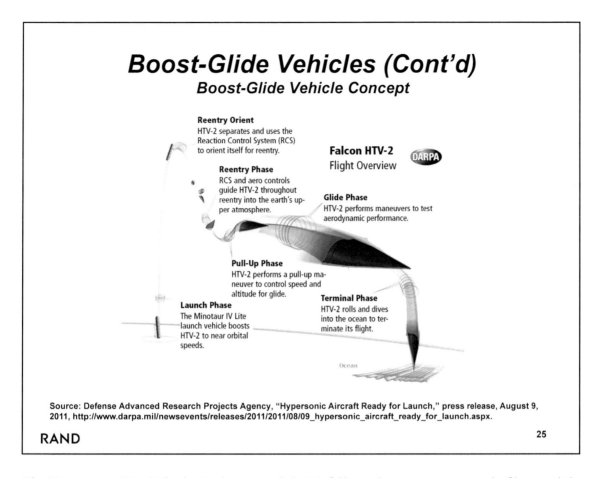

Boost-Glide Vehicles (Cont'd)
Boost-Glide Vehicle Concept

Reentry Orient
HTV-2 separates and uses the Reaction Control System (RCS) to orient itself for reentry.

Reentry Phase
RCS and aero controls guide HTV-2 throughout reentry into the earth's upper atmosphere.

Falcon HTV-2
Flight Overview   DARPA

Glide Phase
HTV-2 performs maneuvers to test aerodynamic performance.

Pull-Up Phase
HTV-2 performs a pull-up maneuver to control speed and altitude for glide.

Launch Phase
The Minotaur IV Lite launch vehicle boosts HTV-2 to near orbital speeds.

Terminal Phase
HTV-2 rolls and dives into the ocean to terminate its flight.

Ocean

Source: Defense Advanced Research Projects Agency, "Hypersonic Aircraft Ready for Launch," press release, August 9, 2011, http://www.darpa.mil/newsevents/releases/2011/2011/08/09_hypersonic_aircraft_ready_for_launch.aspx.

RAND   25

The Hypersonic Test Vehicle–2, shown in slide 25, follows the trajectory typical of boost-glide vehicles.

CHAPTER FOUR

Items Proposed for Category I but Possible Inclusions in Category II

Category I or Category II?

- **Recommendations listed in separate sections**

- **Reluctance to overload Category I**

- **Option to include in Item 5 (currently an open item)**

- **Include items in both Categories I and II?**

RAND   26

This chapter discusses ten items that are specifically applicable to penaids and, therefore, proposed for inclusion in Category I, Item 2.

There may be a reluctance to widen the strict restrictions of Category I to a large number of additional items. For that reason, this report acknowledges the possibility of placing the items under the case-by-case review provisions of Category II. Because these ten items are lower-level subsystems than the three items discussed in the previous chapter, they could arguably be placed into either category. For each of these ten items, we identify possible locations in Category II. An obvious possibility is to place many of them in Category II, Item 5. After a reorganization of the MTCR Annex, some items became open; Item 5 is the first class of items to be open.

The MTCR Annex includes different versions of some items in both Categories I and II. Item 19.A.1 (Category II) specifies rocket systems of any payload with a range capability of 300 km. Subsystems of such rockets—individual rocket stages and rocket motors or engines—are covered in Item 20.A.1. All of these items are downsized versions of Category I items. They are included to control lower-payload missile variants that could deliver chemical or biological agents. Similarly, some items proposed in this report for Category I, such as complete penaid subsystems or missile defense test targets, could be "mirrored" in the Category II controls when they are usable in rocket systems with less than a 500 kg payload capability but with a 300 km range capability.

Canisters/Dispensers

- **Penetration aid packing and/or deployment mechanisms**

- **Issues**
  - **Include in Item 2.A.1.b (re-entry vehicles and equipment designed or modified therefor)?**
  - **Not a complete penaid; it might be deemed less sensitive because it must be combined with other items to create a complete penaid. Include in Item 5 (open item) to avoid overloading Category I?**

RAND   27

Below the level of complete countermeasure subsystems or test targets, there are some new possibilities for MTCR controls. One is to include the new items in Category I, but as a subset of the RV items. This would allow a number of penaid-related items to be added to the Category I list without greatly enlarging the main number of technology classes covered by Category I.

Another possibility is to avoid overloading Category I by creating a new Category II penaid item in the MTCR Annex, using one of the item classes reserved for future use.

Because penaids must be carefully deployed in space, their canisters or dispensers are sophisticated items that warrant inclusion somewhere in the annex.

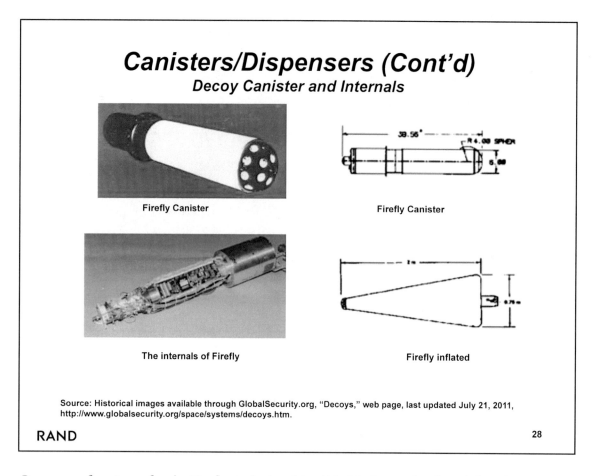

One type of canister, for the Firefly, is depicted in slide 28. It was developed decades ago.

Post-Boost Subsystems

- **Post-boost subsystems providing attitude control or maneuvering capabilities and specially designed for RVs or countermeasure subsystems**

- **Issues**
    - **Without "specially designed," how to distinguish from satellite systems?**
    - **Include in Item 2.A.1.a (individual rocket stages)?**
    - **Include in Item 2.A.1.b (re-entry vehicles and equipment designed or modified therefor)?**
    - **Include in Item 5 (open item) or Item 10 (flight control) to avoid overloading Category I?**

RAND   **29**

In conjunction with canisters and dispensers, post-boost subsystems, which contain small propulsion subsystems, are used to adjust the velocity vector of RVs and penaids.

The problem is that such subsystems can also be used to deploy satellites. The controversial term *specially designed* may need to be used to specify post-boost subsystems uniquely designed for RVs and penaids.

Such subsystems could be included under either of two possible Category I items on the MTCR Annex, or they could be included in two possible places in Category II.

Post-Boost Subsystems (Cont'd)
Post-Boost Subsystem

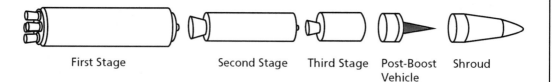

First Stage   Second Stage   Third Stage   Post-Boost   Shroud
                                                       Vehicle

Source: Adapted from Karl Tate, "How Intercontinental Ballistic Missiles Work (Infographic)," Space.com, February 1, 2013, http://www.space.com/19601-how-intercontinental-ballistic-missiles-work-infographic.html.

RAND   **30**

A post-boost vehicle, sometimes called a stage, tends to be a short cylinder containing low-thrust rocket engines and placed between the missile's final boosting stage and the re-entry vehicle(s) and penaid package.

Replicas/Decoys

• **RV replicas or decoys**

• **<u>Issue</u>**

    – **Include in Item 2.A.1.b (re-entry vehicles and equipment designed or modified therefor)?**

    – **Include in Item 5 (open item) to avoid overloading Category I?**

RAND　　　　　　　　　　　　　　31

Items designed to appear to be RVs can be developed as test targets (see slide 22) or developed uniquely as decoy countermeasures.

Electronic Countermeasures

- **Powered devices for electronic warfare (EW) against missile defense systems in any part of the electromagnetic spectrum — including electromagnetic jammers or signal spoofers, anti-jam subsystems, high-power microwave sources, and non-nuclear electromagnetic pulse sources capable of operating in space or re-entry environments**

- **<u>Issues</u>**

    - **Too specific?**

    - **Include in Item 2.A.1.b (re-entry vehicles and equipment designed or modified therefor)?**

    - **Include in Item 5 (open item) or Item 11 (avionics) to avoid overloading Category I?**

RAND   **32**

A variety of electronic countermeasures are appropriate for inclusion in the MTCR Annex. There is a fine line to be walked to avoid being too specific about such items. However, it is important to control penaids over the entire electromagnetic spectrum, as well as both items that generate broad blasts of electromagnetic energy and items that generate finely tuned signals to spoof missile defenses.

Electronic Countermeasures (Cont'd)

Directional Infrared Countermeasures (DIRCMs)

The DIRCM subsystem uses lasers to jam a missile infrared seeker through its aperture. The intended effect is to confuse the missile guidance and cause it to veer off its intercept trajectory.

J-Music DIRCM Subsystem
(Height less than 30 cm)

Source: Promotional product image from Elbit Systems.

RAND   33

DIRCMs can be used on aircraft or missiles to protect them from attack by weapons employing infrared seekers. If the DIRCMs are space-qualified, they could be included in Category I.

Electronic Countermeasures (Cont'd)

Digital Radio Frequency Memory (DRFM) Jammer

DRFM jammers are used to jam or spoof missile defense radars. They are designed to digitize an intercepted radio-frequency signal, then retransmit it to jam the transmitting radar, create a false target, or modify the target characteristics, thereby reducing the effectiveness of the missile defense system.

ADEP-800/1 DRFM Jammer
(Length less than 30 cm)

Source: Promotional product image from Systems and Processes Engineering Corporation, "ADEP™-800/1 DRFM Jammers Delivered to U.S. Army by Systems and Processes Engineering Corporation," press release, July 28, 2009, http://www.spec.com/content/view/68.

RAND   34

Space-qualified DRFM jammers can be used to counter the radars of missile defense systems.

Electronic Countermeasures (Cont'd)

Low-Power Jammer

"... missile attackers can employ low-power jammers (right), pieces of hardware not much larger than a dime, to generate tens of thousands of false targets to mask a warhead's presence or location."

Source: Quote and historical photo from George N. Lewis and and Theodore A. Postol, "The European Missile Defense Folly," *Bulletin of Atomic Scientists*, Vol. 64, No. 2, May 2008.

RAND   35

Large numbers of electronic penaids can be carried in a relatively small payload.

Electronic Countermeasures (Cont'd)
Ukrainian Electromagnetic Pulse and High-Power Microwave Generators

"Magnetocumulative generators (MCG) of electric pulses are based on conversion of explosion energy into that of electric pulses. Owing to the high energy capacity of modern explosives, the magnetocumulative generators of today have pretty small dimensions and low weight and are capable of generating electric pulses that carry currents as high as hundreds of mega-amperes and produce energies up to 100 MJ, their power being as high as 10^{13} W."

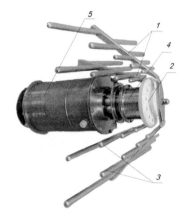

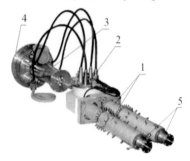

General view of high-power spiral MCG
(Length approximately 0.5 m)

General view of beamless coaxial generator MG1
with connected capacitor storage
(Length less than 0.5 m)

"High-power pulsed beamless microwave generator MG-1 is designed for electromagnetic compatibility (EMC) studies in the laboratory and field conditions."

Source: Institute for Electromagnetic Research, http://iemr.com.ua (click "Products"). Used with permission.

RAND   **36**

A Ukrainian firm, the Institute for Electromagnetic Research, develops items incorporating possible penaid technology. The institute's website does not specifically cite countermeasures as applications for these items.

Chaff/Obscurants/Flares

- **Chaff, obscurants, or flares, and their dispensers, designed to operate in space or re-entry environments, including**
    - **Flares, containing both fuel and oxidizer, with space-qualified igniters**
    - **Passive radio frequency or optical chaff**

- **Issues**
    - **Specify "dispensers" separately from canister/dispenser item above?**
    - **Include in Item 2.A.1.b (re-entry vehicles and equipment designed or modified therefor)?**
    - **Include in Item 5 (open item) to avoid overloading Category I?**

RAND   37

Various kinds of obscurants can generate optical or radio frequency signals to hide an RV. Obscurants require specialized dispensers.

Chaff/Obscurants/Flares (Cont'd)
Ukrainian Chaff and Chaff Dispensers

Photographic image of one of the types of dedicated shape radiolocation reflectors

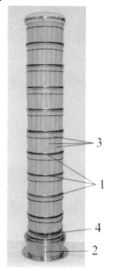

General view of mechanical system of dispersal of dedicated shape radiolocation reflectors

Source: Institute for Electromagnetic Research, http://iemr.com.ua (click "Products"). Used with permission.

RAND   38

As indicated in slide 38, the Institute for Electromagnetic Research has developed chaff for use in the upper or lower atmosphere, as well as associated dispensers.

RV/Decoy Signature Control

- **Mechanisms or specially designed materials for the control of RV or decoy signatures**

- **<u>Issues</u>**
  - **"Specially designed" necessary for "materials"?**
  - **Include "mechanisms" in Item 2.A.1.b (re-entry vehicles and equipment designed or modified therefor)?**
  - **Are materials already adequately covered in Item 17.C.1 (materials for reduced observables)?**
  - **Include in Item 5 (open item) to avoid overloading Category I?**

RAND   **39**

An RV or a decoy can have its signature adjusted to hide it or to increase its visibility in various spectra.

Mechanisms to accomplish this are obvious candidates for the MTCR Annex. However, materials are a more difficult matter. Materials are often dual-use. Therefore, it may be necessary to introduce "specially designed" into their description. If the materials are to be treated as Category II, they could be included under the annex's current provision for stealthy materials. This is Item 17.C.1, "materials for reduced observables such as radar reflectivity, ultraviolet/infrared signatures and acoustic signatures (i.e., stealth technology)." The MTCR makes an exception for "coatings (including paints) when specially used for thermal control of satellites." This item does not currently include materials for *increased* observables.

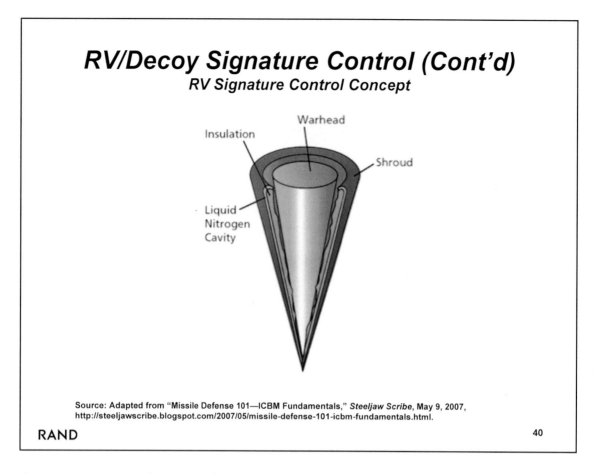

RV/Decoy Signature Control (Cont'd)
RV Signature Control Concept

Source: Adapted from "Missile Defense 101—ICBM Fundamentals," *Steeljaw Scribe*, May 9, 2007, http://steeljawscribe.blogspot.com/2007/05/missile-defense-101-icbm-fundamentals.html.

RAND

40

One signature-control concept, shown in slide 40, would lower the temperature of the RV surface to reduce its infrared visibility.

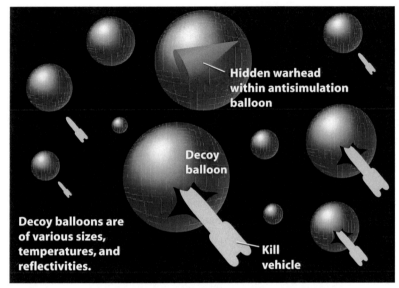

RV/Decoy Signature Control (Cont'd)

Antisimulation Concept

Hidden warhead within antisimulation balloon

Decoy balloon

Decoy balloons are of various sizes, temperatures, and reflectivities.

Kill vehicle

Source: Adapted from Missile Defense Agency illustrations by Alfred T. Kamajian, in Richard L. Garwin, "Holes in the Missile Shield," *Scientific American*, Vol. 291, No. 5, November 2004, http://www.nature.com/scientificamerican/journal/v291/n5/pdf/scientificamerican1104-70.pdf.

RAND

41

Another signature-control concept would disguise the RV to look like a nonlethal object, such as a decoy.

Plume Signature Control

- **Mechanisms or specially designed materials to control the signature of propellant plumes**

- **<u>Issues</u>**
  - **– "Specially designed" necessary for "materials"?**
  - **– Include in Item 5 (open item) to avoid overloading Category I?**

RAND   **42**

Countermeasures can also include efforts to obscure the signature of rocket propellant plumes. Again, there is the issue of whether to use "specially designed" with regard to materials.

Wake Modification

• **Mechanisms or specially designed materials for the control of RV wake signatures**

• <u>**Issues**</u>

    – **"Specially designed" necessary for "materials"?**

    – **Include in Item 2.A.1.b (re-entry vehicles and equipment designed or modified therefor)?**

    – **Include in Item 5 (open item) to avoid overloading Category I?**

RAND   **43**

On re-entry, the wake of an RV can reveal its location, enabling it to be targeted by terminal missile defenses. The inclusion of wake modification items in the MTCR Annex would raise the usual "specially designed" problem of how to handle materials.

Maneuvering Subsystems

- **Specially designed subsystems for RVs that enable maneuvering**

- **Issues**
    - **"Specially designed" necessary to avoid subsystems for peaceful spacecraft?**
    - **Include in Item 2.A.1.b (re-entry vehicles and equipment designed or modified therefor)?**
    - **Include in Item 5 (open item) or Item 10 (flight control) to avoid overloading Category I?**

RAND   **44**

Maneuvering RVs—exoatmospheric or endoatmospheric—can evade missile defenses. However, thrusted, exoatmospheric maneuvering subsystems are also used for the flight control of peaceful spacecraft. Thus, "specially designed" again becomes an issue. Likewise, aerodynamic and/or endoatmospheric maneuvering systems would seem to raise the "specially designed" issue. Note that complete RVs are already included in Category 1 of the annex, so this item applies only to subsystems for RVs.

Submunitions

- **Missile submunitions specially designed for the delivery of liquids, gases, or powders**

Issues

- – **"For the delivery of liquids, gases, or powders" necessary for chemical or biological agents?**
- – **Include in Item 2.A.1.b (re-entry vehicles and equipment designed or modified therefor)?**
- – **Leave for chemical/biological nonproliferation regimes?**
- – **Include in Item 5 (open item)?**

RAND   **45**

Submunitions, large numbers of small space-qualified payloads released from a missile, can serve as penaids by multiplying the number of targets confronting a missile defense. Because the MTCR is explicitly directed against the proliferation of missiles capable of delivering WMD, it could logically restrict submunitions capable of delivering chemical or biological agents.[1]

However, a submunition subsystem for chemical or biological delivery is arguably already restricted by international export controls on chemical and biological weapons. Given this consideration, submunitions might not need to be added to the MTCR.

There are also submunitions for the delivery of conventional, high-explosive payloads for cratering runways or large-area attacks against personnel or vehicles. However, these fall outside the realm of WMD and, therefore, outside the controls of the MTCR. They might, however, be covered by the international Wassenaar Arrangement, which covers exports of conventional weapons but is less restrictive than the MTCR.

---

[1] Nuclear warheads may be delivered by multiple re-entry vehicles, but they are generally regarded as too massive to be released in very large numbers (i.e., more than 15 or so) from a single missile. Missiles for delivering radiological agents are not explicitly targeted by the MTCR guidelines.

Submunitions (Cont'd)

U.S. cluster warhead, circa 1960

Sources: U.S. Army photo, in "Demonstration Cluster Bomb," *Wikipedia*, last updated March 27, 2009, http://en.wikipedia.org/wiki/File:Demonstration_cluster_bomb.jpg.

RAND   46

Slide 46 illustrates chemical submunitions carried by a short-range U.S. missile in the 1960 period. As a result of international arms-control undertakings, the United States no longer possesses biological weapons and is destroying its chemical weapon stocks.

CHAPTER FIVE

Items Proposed for Category II

Penaids: Possible Category II Items

These items have alternative uses that should not necessarily be as tightly restricted as Category I items. The items can be used for satellites or missile defenses themselves.

RAND   47

The items addressed in this chapter are clearly dual-use and therefore appropriate for Category II.

Multiple-Object Deployment

- **Items, other than the specially designed post-boost subsystems described for Category I, for the automatic or remotely controlled ejection or deployment of multiple flight objects in space except when incorporated in satellites**

- **<u>Issues</u>**
  - **How to distinguish from penaid post-boost subsystems?**
  - **Exempt "when incorporated in satellites" to conform to other MTCR exemptions?**
  - **Include in Item 5 (open item) or Item 10 (flight control)?**

RAND   **48**

Penaids can consist of many objects, including decoys, jammers, chaff, and flares. Special mechanisms, such as canisters attached to post-boost vehicles, deploy these objects appropriately. However, similar mechanisms are used to deploy multiple satellites from a single rocket. Consideration must thus be given to exempting mechanisms that are not "specially designed" as penaid post-boost vehicles. The MTCR already includes exemptions for satellite items, and this type of exemption might be used here.

Inflation/Assembly Items

- **Items for the automatic or remotely controlled inflation or assembly of objects in space, except when incorporated in satellites**

- **<u>Issues</u>**
  - **Exempt "when incorporated in satellites" to conform to other MTCR exemptions?**
  - **Include in Item 5 (open item)?**

RAND   **49**

Test targets can be balloons or decoys that require inflation or other forms of automatic assembly in space. However, some satellites are balloons or otherwise involve automatic assembly, such as for solar panels. So, a satellite exemption may be appropriate.

Inflation/Assembly Items (Cont'd)
NASA Inflatable Re-Entry Shield

Source: NASA photo by Kathy Barnsdorf, in Leonard David, "NASA Launching High-Tech Inflatable Heat Shield Test Monday," *Space.com*, July 17, 2012, http://www.space.com/16615-nasa-inflatable-heat-shield-launching-saturday.html.

RAND   50

A recent example of an inflated space object is a National Aeronautics and Space Administration (NASA) re-entry heat shield, tested in July 2012.

Hardening

- **Items specially designed for hardening against thermal radiation, non-nuclear electromagnetic pulses, or high-powered microwaves**

- **Issues**
  - **Include in Item 5 (open item) or Item 18 (nuclear effects protection), which already covers hardening against nuclear radiation pulses and nuclear "combined blast and thermal effects"?**

RAND   51

The MTCR already includes items for hardening RVs against certain types of nuclear effects. However, it may be appropriate to add other forms of hardening against missile defenses.

Attack Warning Sensors

- **Mechanisms to be carried on a missile-borne object to sense when it or a companion object is under attack**

- **Issues**
    - **Include in Category II because item is usable in other systems, such as aircraft?**
    - **Include in Item 5 (open item) or Item 11 (avionics) to avoid overloading Category I?**

RAND   **52**

RVs can wait for warning of missile defenses before deploying countermeasures. The warning systems are candidates for the MTCR Annex. However, such warning systems have applications other than ballistic missile countermeasures, so they are candidates for inclusion in Category II.

Fly-Along Sensors

• **Fly-along sensor packages**

• **<u>Issues</u>**

    – **Item 5 (open item) or Item 15 (test facilities and equipment)?**

RAND

53

One form of instrumentation for testing penaids is the fly-along sensor, which is dispensed with the RV and its penaids and reports to a ground station on the penaids' performance. However, such sensors can also report on satellite performance, thus making a Category II designation appropriate.

Test Facilities/Equipment

- **Test facilities, test equipment, and diagnostic (e.g., measuring and calibration) equipment usable for controlled items, including**
    - **Vacuum chambers suitable for the drop testing of controlled items**
    - **Signature measurement chambers**

- **Issues**
    - **Include in Item 5 (open item) or Item 15 (test facilities and equipment)?**
    - **Distribute among specific items or place in Item 5?**

RAND   **54**

Other types of test instrumentation and facilities are usable for both penaids and satellite development or microgravity research.

Test Facilities/Equipment (Cont'd)
German and U.S. Vacuum Drop Towers

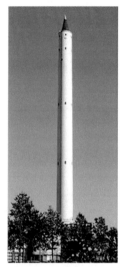

Bremen drop tower

NASA drop tower. View of free fall in vacuum chamber

Sources: "Fallturm Bremen," *Wikipedia*, last updated March 13, 2013, http://en.wikipedia.org/wiki/Fallturm_Bremen; NASA, "Zero Gravity Research Facility," web page, last updated September 28, 2011, https://rt.grc.nasa.gov/main/rlc/zero-gravity-research-facility.

RAND

55

A vacuum drop tower is used for the terrestrial testing of items, allowing them to experience zero gravity for several seconds. Again, such facilities are usable for both penaid development and microgravity research.

CHAPTER SIX

Implementing Penaid Export Controls

Implementation

- ## Review criteria

- ## Assessment of intent

- ## Risk of overloading

- ## Consensus requirements

RAND   **56**

Although penaid export controls present some definitional and structural issues, such issues are familiar matters in the implementation of the MTCR.

For example, the problem of individualizing the decisions on the export of dual-use items is broadly handled by the MTCR's Category II list, which consists of items subject to case-by-case review rather than a strong presumption of export denial. Category II items (and, indeed, all MTCR items) are reviewed by the following six criteria (see MTCR, undated, para. 3):

A. Concerns about the proliferation of weapons of mass destruction;

B. The capabilities and objectives of the missile and space programs of the recipient state;

C. The significance of the transfer in terms of the potential development of delivery systems (other than manned aircraft) for weapons of mass destruction;

D. The assessment of the end use of the transfers, including relevant assurances of the recipient states;

E. The applicability of relevant multilateral agreements;

F. The risk of controlled items falling into the hands of terrorist groups and individuals.

Some of these criteria involve tests of intent (e.g., "objectives," "assessment of the end use," "assurances"). Although many of the current objections to the use of "specially designed" are grounded in a preference for physical descriptions over estimates of intent, the MTCR is replete with the term *specially designed* and with criteria of intent. Similarly, countermeasure equipment usable for both aircraft and missiles (such as chaff and electronic jammers) or usable for both peaceful spacecraft and missiles (such as inflation/assembly items) can be reviewed for export with the above criteria.

Another concern, discussed at length earlier, is the possible negotiating burden if the MTCR—particularly the highly restrictive Category I list—is overloaded with up to 19 new items. However, new items can be nested into the control definitions of larger items. For instance, MTCR Annex Item 2.A.1.b (re-entry vehicles and equipment designed or modified therefor), is not a single item. It already has sub-items: (1) heat shields, (2) heat sinks, and (3) electronic equipment specially designed for re-entry vehicles. More sub-items could be added to this list. Alternatively, a new Category I item 2.A.1.g could be added as "penetration aids" with sub-items suggested in this report for MTCR Item 2.

The greatest difficulty will be obtaining the support of key governments for penaid export controls. The five nations defined by the Nuclear Non-Proliferation Treaty as nuclear-armed states—the United States, the United Kingdom, France, Russia, and China—all have sophisticated penaid programs. Modifications to the MTCR require the consensus of all 34 members, including Russia but not China (which, as noted earlier, professes to observe a different form of the MTCR from the version currently in effect). The Congressional Commission on the Strategic Posture of the United States singled out Russia and China for special attention. Russia's agreement is needed for an MTCR modification, and China's support is needed if there is not to be a serious loophole in penaid export controls.

CHAPTER SEVEN

Concluding Observations

Conclusion

Penaid export controls would help

• **Missile nonproliferation**

• **Missile defense**

• **Deterrence**

But will Russia and China cooperate?

RAND   57

This research illustrates how the MTCR Annex can be modified to provide better controls on missile defense penetration aids. If enacted, the MTCR modifications suggested here would constitute one of the most significant adjustments to the regime since its inception in 1987. The recommended modifications, or some variant of them, would strengthen the regime's ability to impede the spread of increasingly lethal ballistic missiles capable of delivering WMD and penetrating missile defenses.

Although policy considerations were beyond the scope of this research, moving a complex regime modification to fruition would obviously require careful diplomacy by the United States and like-minded governments. Several government officials interviewed for this study believe such an effort would be worthwhile. The recommended MTCR revisions would reinforce the effectiveness of U.S. and allied missile defense systems, in turn bolstering protection and deterrence against missile-armed adversaries and enhancing international security.

The largest outstanding question is not the value of restrictions on penaid exports. It is whether Russia and China will support such restrictions.

References

Burnell, Brian, "Chevaline Deployment Sequence–Mod," *Wikipedia*, last modified December 30, 2011. As of June 28, 2013:
http://en.wikipedia.org/wiki/File:Chevaline_deployment_sequence-mod.gif

"Chevaline," *Wikipedia*, last modified June 24, 2013. As of June 28, 2013:
http://en.wikipedia.org/wiki/Chevaline

Congressional Commission on the Strategic Posture of the United States, *America's Strategic Posture: The Final Report of the Congressional Commission on the Strategic Posture of the United States*, Washington, D.C.: United States Institute of Peace Press, May 2009. As of June 28, 2013:
http://media.usip.org/reports/strat_posture_report.pdf

Davenport, Kelsey, Daniel Horner, and Daryl G. Kimball, "Missile Control: An Interview with Deputy Assistant Secretary of State Vann Van Diepen,"
Arms Control Today, July–August 2012. As of June 28, 2013:
http://www.armscontrol.org/2012_07-08/Interview_With_Deputy_Assistant_Secretary_Of_State_Vann_Van_Diepen

David, Leonard, "NASA Launching High-Tech Inflatable Heat Shield Test Monday," Space.com, July 17, 2012. As of June 28, 2013:
http://www.space.com/16615-nasa-inflatable-heat-shield-launching-saturday.html

Defense Advanced Research Projects Agency, "Hypersonic Aircraft Ready for Launch," press release, August 9, 2011. As of June 28, 2013:
http://www.darpa.mil/newsevents/releases/2011/2011/08/09_hypersonic_aircraft_ready_for_launch.aspx

"Demonstration Cluster Bomb," *Wikipedia*, last updated March 27, 2009. As of June 28, 2013:
http://en.wikipedia.org/wiki/File:Demonstration_cluster_bomb.jpg

Dommett, R. L. "Engineering the Chevaline Delivery System," *Prospero: Proceedings from the British Rocket Oral History Conferences at Charterhouse*, No. 5, Spring 2008.

"Fallturm Bremen," *Wikipedia*, last updated March 13, 2013. As of June 28, 2013:
http://en.wikipedia.org/wiki/Fallturm_Bremen

GAO—*See* U.S. Government Accountability Office.

Garwin, Richard L., "Holes in the Missile Shield," *Scientific American*, Vol. 291, No. 5, November 2004, pp. 70–79. As of June 28, 2013:
http://www.nature.com/scientificamerican/journal/v291/n5/pdf/scientificamerican1104-70.pdf

GlobalSecurity.org, "Decoys," web page, last updated July 21, 2011. As of June 28, 2013:
http://www.globalsecurity.org/space/systems/decoys.htm

Institute for Electromagnetic Research, Ltd., "Beamless Pulsed Microwave Generators," web page, undated(a). As of June 28, 2013 (click "Products"):
http://iemr.com.ua

———, "Chaffs of Complicated Form," web page, undated(b). As of June 28, 2013 (click "Products"):
http://iemr.com.ua

———, "Dispensors of Chaffs of Complicated Form," web page, undated(c). As of June 28, 2013 (click "Products"):
http://iemr.com.ua

———, "Magnetocumulative Generators of High-Power Electric Pulses,"web page, undated(d). As of June 28, 2013 (click "Products"):
http://iemr.com.ua

Jones, Peter, "Chevaline Technical Programme 1966–1976," *Prospero: Proceedings from the British Rocket Oral History Conferences at Charterhouse*, No. 2, Spring 2005.

Lewis, George N., and Theodore A. Postol, "The European Missile Defense Folly," *Bulletin of Atomic Scientists*, Vol. 64, No. 2, May 2008, pp. 32–39.

L'Garde, Inc., "The Calibrated Orbiting Objects Project (COOP)," web page, undated(a). As of June 28, 2013:
http://www.lgarde.com/programs/missile-defense-targets-and-countermeasures/coop

———, "Missile Defense Targets and Countermeasures," web page, undated(b). As of June 28, 2013:
http://www.lgarde.com/programs/missile-defense-targets-and-countermeasures

McMahon, K. Scott, Stanley Orman, and Richard Speier, "Penaid Nonproliferation: Analysis and Recommendations," briefing, Veridian Corp., Arlington, Va., June 29, 2000.

"Missile Defense 101—ICBM Fundamentals," *Steeljaw Scribe*, May 9, 2007. As of June 28, 2013:
http://steeljawscribe.blogspot.com/2007/05/missile-defense-101-icbm-fundamentals.html

Missile Technology Control Regime, "Guidelines for Sensitive Missile-Relevant Transfers," undated. As of June 28, 2013:
http://www.mtcr.info/english/guidetext.htm

———, "Equipment, Software and Technology Annex," October 23, 2012. As of June 28, 2013:
http://www.mtcr.info/english/annex.html

MTCR—*See* Missile Technology Control Regime.

NASA—*See* National Aeronautics and Space Administration.

National Aeronautics and Space Administration, Glenn Research Center, "Zero Gravity Research Facility," web page, last updated September 28, 2011. As of June 28, 2013:
https://rt.grc.nasa.gov/main/rlc/zero-gravity-research-facility

Panton, Frank, "Polaris Improvements and the Chevaline System 1967–1975/6," *Prospero: Proceedings from the British Rocket Oral History Conferences at Charterhouse*, No. 1, Spring 2004a.

———, "The Unveiling of Chevaline: House of Commons Public Accounts Committee 1981–2," *Prospero: Proceedings from the British Rocket Oral History Conferences at Charterhouse*, No. 1, Spring 2004b.

———, "Polaris Improvement Programme: Chevaline System," *Prospero: Proceedings from the British Rocket Oral History Conferences at Charterhouse*, No. 3 , Spring 2006.

"Penetration Aid," *Wikipedia*, last updated March 17, 2013. As of June 28, 2013:
http://en.wikipedia.org/wiki/Penetration_aid

Sessler, Andrew M., John M. Cornwall, Bob Dietz, Steve Fetter, Sherman Frankel, Richard L. Garwin, Kurt Gottfried, Lisbeth Gronlund, George N. Lewis, Theodore A. Postol, and David C. Wright, *Countermeasures: A Technical Evaluation of the Operational Effectiveness of the Planned US National Missile Defense System*, Cambridge, Mass.: Union of Concerned Scientists, April 2000. As of June 28, 2013:
http://www.ucsusa.org/nuclear_weapons_and_global_security/missile_defense/technical_issues/countermeasures-a-technical.html

Simpson, John, "The History of the UK Strategic Deterrent: The Chevaline Programme," *Proceedings of the Royal Aeronautic Society*, London, October 28, 2004.

Speier, Richard, "The Missile Technology Control Regime," in Trevor Findlay, ed., *Chemical Weapons and Missile Proliferation*, Boulder, Colo.: Lynne Rienner Publishers, 1991.

———, *The Missile Technology Control Regime: Case Study of a Multilateral Negotiation*, Washington, D.C.: United States Institute of Peace, November 1995.

————, *Compendium of Non Proliferation Documents*, volume of the study *Arms Control Impact on Target Missiles for Defense Testing*, McLean, Va.: JAYCOR, February 2000.

————, "Missile Nonproliferation and Missile Defense: Fitting Them Together," *Arms Control Today*, November 2007. As of June 28, 2013:
http://www.armscontrol.org/act/2007_11/Speier

Speier, Richard H., Brian G. Chow, and S. Rae Starr, *Nonproliferation Sanctions*, Santa Monica, Calif.: RAND Corporation, MR-1285-OSD, 2001. As of June 28, 2013:
http://www.rand.org/pubs/monograph_reports/MR1285.html

Systems and Processes Engineering Corporation, "ADEP™-800/1 DRFM Jammers Delivered to U.S. Army by Systems and Processes Engineering Corporation," press release, July 28, 2009. As of June 28, 2013:
http://www.spec.com/content/view/68

Tate, Karl, "How Intercontinental Ballistic Missiles Work (Infographic)," Space.com, February 1, 2013. As of June 28, 2013:
http://www.space.com/19601-how-intercontinental-ballistic-missiles-work-infographic.html

U.S. Government Accountability Office, *Defense Acquisitions: Sound Business Case Needed to Implement Missile Defense Agency's Targets Program*, Washington, D.C., GAO-08-1113, September 2008.

————, *Defense Acquisitions: Assessments of Selected Weapon Programs*, Washington, D.C., GAO-11-233SP, March 2011.